PENGUINS
IN THE WILD

A Visual Essay

JOE MCDONALD

AMHERST MEDIA, INC. ■ BUFFALO, NY

Praise for *Penguins in the Wild:*

"Joe and Mary Ann McDonald have compiled a visual and educational discovery of the penguin world that's illustrated with world-class photography and fascinating to read."

—Susan Day, Executive Director of the North American Nature Photography Association.

Copyright © 2019 by Joe McDonald.
All rights reserved.
All photographs by Joe McDonald and Mary Ann McDonald unless otherwise noted.

Published by:
Amherst Media, Inc.
PO BOX 538
Buffalo, NY 14213
www.AmherstMedia.com

Publisher: Craig Alesse
Senior Editor/Production Manager: Michelle Perkins
Editors: Barbara A. Lynch-Johnt, Beth Alesse
Acquisitions Editor: Harvey Goldstein
Associate Publisher: Katie Kiss
Editorial Assistance from: Carey A. Miller, Roy Bakos, Jen Sexton-Riley, Rebecca Rudell
Business Manager: Sarah Loder
Marketing Associate: Tonya Flickinger

ISBN-13: 978-1-68203-372-2
Library of Congress Control Number: 2018936010
Printed in the United States of America
10 9 8 7 6 5 4 3 2 1

AUTHOR A BOOK WITH AMHERST MEDIA!

Are you an accomplished photographer with devoted fans? Consider authoring a book with us and share your quality images and wisdom with your fans. It's a great way to build your business and brand through a high-quality, full-color printed book sold worldwide. Our experienced team makes it easy and rewarding for each book sold—no cost to you. E-mail **submissions@amherstmedia.com** *today!*

www.facebook.com/AmherstMediaInc
www.youtube.com/AmherstMedia
www.twitter.com/AmherstMedia
www.instagram.com/amherstmediaphotobooks

CONTENTS

Acknowledgments

Working on this book and making the photographs would not have been possible without the invaluable help of many people. First off, I must thank Adam, Lisle, Angus, Joe, Cindy, and Donna for their wonderful photographs, rounding out a penguin profile that truly reveals these birds' incredible lives. Equipment is the life-blood of a photographer, so I must thank Joe Johnson and Jim Weise from Really Right Stuff, Lou Schmidt from Hoodman, Ken Hubbard from Tamron, Gary Farber from Hunt's Photo, Paul DeZeeuw from Cognisys, and Clay Wimberley from Wimberley Tripod Heads for all their past support, for this project and for so many more. All make or sell top-quality equipment that makes Mary's and my photography so much more productive. Every photograph in this book made by Mary or me was shot in the company of our friends on our photo tours and safaris, and we extend special thanks to our good friends Bill Sailer, Tom Wester, Judy Johnson, and Don Lewis, who have shared so many adventures with Mary and me. Last, our visits to Antarctica and South Georgia would not be possible if we were not co-leading these photo safaris with our good friend and chief competitor in the photo tour business, Joe Van Os. Mary and I have shared many adventures in the deep South with Joe and our fellow leaders. Last, my deepest appreciation and love goes to the person who helps make our exciting life possible—my wife, Mary Ann, the focus of my world.

ABOUT JOE AND MARY ANN MCDONALD

Joe McDonald. Joe McDonald has been a full-time professional wildlife photographer since 1983, and is the author of 16 books on wildlife and wildlife photography. He has spent months in penguin country, including numerous trips to Antarctica, South Georgia, the Falklands, and the Galápagos. When he is not photographing penguins, Joe is traveling the world, leading photo safaris to Africa, India, South America, and elsewhere.

Mary Ann McDonald. Mary Ann McDonald is the author of 29 children's books on natural history, ranging in topic from Jupiter to chickens, mosquitoes to pythons. Mary co-leads the photo expeditions, tours, and safaris with her husband, Joe, and has been a full-time photographer since 1999. In addition to her duties as a photographer and tour leader and teacher, Mary runs the business end of their busy life.

Other Photo Contributors

Adam Rheborg. Switzerland-born Adam Rheborg is a worldwide explorer who has spent the past two decades working as an expedition leader and guide, a scuba instructor, and a photographer and lecturer in the polar regions, as well as in South America, Africa, and Asia. Adam is the expedition leader on the McDonalds' trips to Svalbard for polar bears and other wildlife.

Lisle Gwynn. British-born Lisle Gwynn is a professional bird-watching guide, leading trips around the world for bird-watching and wildlife photography. Lisle has guided photo tours to the Russian far east, the South Pacific and New Zealand, southern Africa, and several South American destinations. He is the tour guide for the McDonald's Ecuador and South Africa photo safaris.

Angus Fraser. Australian Angus Fraser is a professional biologist and extremely energetic and serious wildlife photographer. Angus has traveled widely, including photo expeditions for snow leopards, jaguars, pumas, and, most recently, the Galápagos Islands where he photographed Galápagos penguins, snorkeling far deeper than Joe or Mary Ann. He is a personal friend.

Joe Van Os. Our friend, Joe Van Os, operates Joseph Van Os Photo Safaris, and it is with Joe that Mary and I have traveled to South Georgia and to Antarctica. Joe has visited many emperor penguin colonies, and was once marooned at a colony for several weeks because of bad weather.

Cindy Marple. Cindy Marple is a retired engineer and very serious bird watcher and photographer. She has traveled to New Zealand several times, as well as to South Georgia, Antarctica, and the Falklands to photograph penguins and other seabirds. Cindy lives in Arizona and is a good friend.

Donna Salett. Donna Salett lives in Massachusetts and is a serious wildlife photographer. She has traveled with the McDonalds to India and Africa, and Donna has extensive coverage of African penguins. Donna and her husband Ron are our good friends.

INTRODUCTION

The penguin. Perhaps the most recognized type of bird in the entire world, and one that brings smiles to all who sees one. Standing erect, dressed in a coat of feathers that resembles a man's tuxedo, waddling along like a toddler taking his or her first steps, the penguin seems almost human, and certainly reminds us of ourselves. The 17 to 20 different species are scattered across the southern hemisphere, a remarkable achievement for a relatively small group of similar-looking birds. The penguin conjures up images of cold, ice-bound lands—the one animal virtually everyone would associate with the continent of Antarctica.

While 5 species of penguin are found in Antarctica, only two are found there exclusively. One, the cute and animated Adélie penguin, has achieved some form of immortality from films like *Happy Feet*. The other is perhaps less well known, but equally famous as the stars of another feature film, *March of the Penguins,* that covered the life of the world's largest penguin, appropriately named the emperor penguin. All of the other species have ranges that extend far beyond the inhospitable world of the southernmost continent. One, in fact, lives on the equator, on a remote and barren archipelago nearly 600 miles off the coast of Ecuador, South America. Here, on the islands the Galápagos, penguins share the cold waters of the Humboldt Current with otherwise tropical wildlife: frigatebirds, pelicans, boobies, marine iguanas, and green sea turtles.

Between the two extremes, the remaining 17 species live, dwelling along the cold coastlines of western South America, southern Africa, southern Australia and New Zealand, and scores of isolated islands spanning the lower half of the southern oceans. All penguins are flightless and live the majority of their life at sea, but like all birds, penguins must return to land to lay their eggs in some type of nest or burrow. But there is one exception, and that is the emperor penguin, whose entire life cycle revolves around either water or ice. This bird doesn't build a nest or claim a territory; instead, this amazing bird forms nesting colonies of hundreds or thousands of birds on the

> "All penguins are flightless and live the majority of their life at sea, but like all birds, penguins must return to land to lay their eggs in some type of nest or burrow."

Emperor Penguin. Photo by Adam Rheborg. Although the penguins are a small family of birds, with 17 to 20 species recognized, they are among the most widespread, with species living below the Antarctic circle, like these emperor penguins, and others living in the tropics.

Antarctic ice shelves. Instead of coddling its egg or young chick in a pebble-lined nest or a warm burrow, emperors incubate and rear their chick on their webbed feet, an inch or so above the killing cold of their ice-covered landscape.

With this one exception, all other penguins establish colonies or nesting sites on land, sometimes hundreds of yards from the ocean shores. Several dig burrows, using the sharp claws of their webbed feet to dig, and then paddling the loose soil to the burrow entrance with both their flippers and feet. Others build miniature hillocks of stone or pebbles, forming crater-like cavities in the center to cradle their two eggs or tiny chicks. No penguin nests in trees, but some species in New Zealand live in the thick temperate forests, in a fairy land of twisted, moss-covered rhododendron and ancient ferns.

While all penguins superficially look alike with their white bellies and dark backs, the penguin clan is divided into six distinct groups or genera, representing 17 to 20 different species. Scientists are not in agreement here, as some suggest a few subspecies should be "split" into their own species designation. It can be confusing, and in this book we'll deal with the group as 17 species, as the 3 in question are quite similar. Emperors, and their close look-alike, the king, comprise one of these groups, the genus Aptenodytes, commonly known as the large penguins, because of their size. Male emperor penguins can stand nearly 4 feet tall! Another genus, comprised of the most members with 6 (or 8) species, are the crested penguins, the genus Eudyptes.

> "Unlike all other species, the little penguins are nocturnal and travel to and from the sea under the cover of darkness."

Two genera have only one or two members. One, the Eudyptula, has the smallest species, although scientists are not in total agreement as to whether the white-flippered penguin is simply a subspecies of the little penguin or is a different species in the same genus. Regardless, standing around 15 inches high, these tiny penguins are less than a third the size of their largest cousin, the emperor. Unlike all other species, the little penguins are nocturnal and travel to and from the sea under the cover of darkness. One genus has but a single representative, the yellow-eyed penguin, so named for its unique yellow eyes. Banded penguins include the Galápagos penguin and 3 other species living on the continent of South America and Africa. The sixth group are called brush-tailed penguins, as all 3 species have stiff, bristly tail feathers that form the third leg of a supportive tripod-like base when the birds are standing.

Seeing any of these penguins in person, in the wild, provides an entirely different picture, as it were, from that derived from photographs, where the birds' plumage gleams white and spotless, and the birds are mute. In life, penguins are often noisy and raucous, and in some large colonies there is a continuous undertone of whistles or trumpets or wispy high-pitched cries. At some nesting colonies, where birds freely excrete in every direction long pinkish-red streams, a sure sign of a diet of krill, birds

Galápagos Penguin. The Galápagos penguin is the northernmost species, reaching all the way to the equator. In their fishing forays, some actually wander into the northern hemisphere, at least for a few miles.

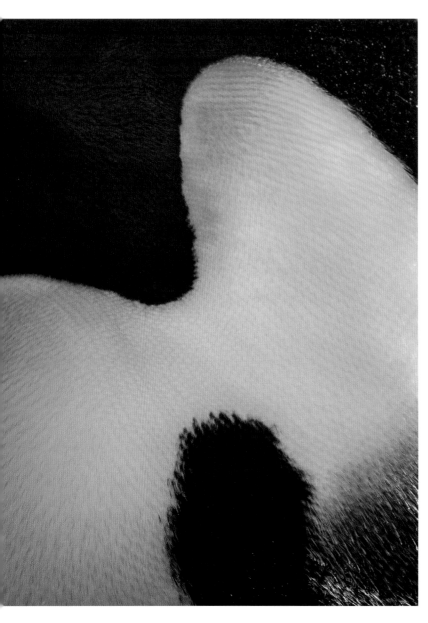

Emperor Penguin. Photo by Adam Rheborg. The simplicity of a penguin's color pattern is nonetheless strikingly beautiful, as this close-up of an emperor penguin clearly shows.

the sea. Their flippers are modified wings, with fused, solid bones perfect for propulsion in the thick medium of water, allowing the birds to fly underwater, flapping their solid flippers much as a bird's wings would do so in the air. One could almost say penguins can fly, for when they really want to move at top speed, penguins will porpoise through the sea, repeatedly popping up in a gentle arch as they race to shore, flee a predator, or head out to their hunting grounds.

It is at sea that penguins are among their most vulnerable, where they face their most common natural predators: leopard seals, fur seals, orcas, and sharks. But the sea holds other risks, too. Over-fishing, by man, has drastically reduced the anchovies and sardines African penguins rely on, and now even krill, a mainstay for many Antarctic species, is being harvested for, of all things, food for commercial fish operations and as a vitamin supplement for humans. Climate change also poses grave concerns as this may shift

can be anything but pretty. Sometimes, at these nesting colonies, I've found it almost impossible to find a clean-looking bird at a nest to photograph. Every adult was splattered in some degree with a coating of red sludge.

A dirty-looking nesting bird tending its eggs or chicks in a crowded colony can be forgiven, for penguins are truly birds of ocean currents and alter the ranges of fish and krill stocks. These changes threaten not only penguins but many other species as well, including seals and whales.

Paradoxically, a warmer climate and warmer ocean temperatures can result in more snow on the Antarctic Peninsula, causing nest abandonment as incubating Adélie penguins are buried beneath feet of snow, while in other regions unseasonal rains flood nests and soak downy chicks that may die from hypothermia. The periodic ocean warming from El Niño events threaten Galápagos and Humboldt penguins. Always, there is the threat of oil spills, from tankers or from damaged vessels or from

> "Indeed, two-thirds of the world's penguin species are classified as vulnerable, threatened, or endangered."

off-shore drilling, posing danger not only to sea-going penguins, but to shoreline nesting colonies, as well. Indeed, two-thirds of the world's penguin species are classified as vulnerable, threatened, or endangered.

These depressing facts belie the general impression one may have when one stands before a king penguin colony on South Georgia, where a hundred thousand pairs of birds may stretch from left to right and

King Penguin and Snowy Sheathbill. Although king penguins are built for a life on the sea, they must return to land to breed, where they share the beaches with elephant seals and pesky snowy sheathbills. Here, a sheathbill pecks at a loose piece of skin and feathers.

Chinstrap Penguins. The dark backs of these chinstrap penguins help to camouflage the bird when seen from above, useful when hunting for fish overhead. Seen from below, the white bellies blend with the sky, and this may lessen the chances that a lurking leopard seal will spot an unwary bird.

the dots of white and black of nesting birds carpet the steep hillsides in a nearly overwhelming visual and auditory spectacle. Birds stream past as they return from the sea, their plumage spotless and glistening, their personalities trusting and curious. To stand at such a place, to be inspected by a curious penguin, and to be bombarded by thousands of bird voices evokes such emotion and awe that it is a spectacle rivaled by few others in the natural world. This is the world of the penguin that we will explore.

" ... to be inspected by a curious penguin, and to be bombarded by thousands of bird voices evokes such emotion and awe. . . "

King Penguins. With the air ringing with the whistles and trumpeting of thousands of birds and the smell of guano, an enormous colony of king penguins can be overwhelming to the human visitor—a total sensory experience.

ORIGINS

As odd as it may seem, the name "penguin" once applied to a bird living in the northern hemisphere that was completely unrelated to the southern penguins. This was the great auk, a flightless bird related to the comical-looking puffin, and a bird that was driven to extinction by 1844, exploited unmercifully as an easy source of meat and fresh eggs. The first sailors heading to the

Razorbill Auk. (left) Penguins get their name by mistaken identity. The original "penguin" is the now extinct relative of the modern-day razorbill, a northern hemisphere bird that can fly but sports similar-looking plumage. **Common Murre.** (right) Although they superficially resemble a penguin, common murres are strong fliers, but murres also use their relatively short, powerful wings to "fly" underwater, much like penguins do with their flipper-like wings. Most likely, the ancestor of penguins had similar habits to the murres.

King Penguins Marching. King penguins in a convoy make tracks along the beach. Penguins are the consummate seabird, but they still must return to land to nest.

southern oceans, perhaps familiar with the great auk, named the similarly patterned and flightless birds they discovered there "penguins."

Penguins, in one form or another, have been around for a long time. The oldest known fossils, some 61 million years old, belong to a giant penguin that stood around 50 inches tall, but its development suggests penguins evolved long before that, during the time of the dinosaurs in the Cretaceous period. Later, around 33 million years ago, a giant penguin, Anthropornis nordenskjoeldi, stood nearly 6 feet tall and weighed over 200 pounds! In contrast, the largest species today, the emperor penguin, is just over half that size and only half the weight.

"The oldest known fossils, some 61 million years old, belong to a giant penguin that stood around 50 inches tall ..."

Adélie and Emperor Penguins. Photo by Adam Rheborg. Antarctica and penguins seem to be synonymous, but only two species, the emperor penguin, the largest species, and the Adélie penguin, barely half its size, are restricted to the Antarctic continent and breed nowhere else.

King Penguin's Tongue. All modern birds are toothless, although some species have serrated beaks that resemble teeth. Although all penguins have smooth beaks, they have a bristly looking tongue that helps the bird to secure the fish and krill it captures. This raggedy-looking bird is molting and will remain on land until its old feathers are replaced.

Gentoo Penguin Porpoising. (following page, top) Penguins are truly birds of the sea and use their powerful flippers to propel themselves through the water. When traveling fast, penguins, like this gentoo, porpoise in long series of jumps above the waves.

Black King Penguin. (top left) All penguins are basically bi-colored, with white bellies and a gray or blackish back. This unfortunate King Penguin was mired in thick, sticky black mud that coated its feathers. Luckily, when it returns to the sea, the muck will wash off and the bird's striking, gleaming plumage will once more be revealed. **Royal and King Penguins.** (bottom left) Photo by Lisle Gwynn. All other species of penguin nest in sub-Antarctic and even temperate regions of the Southern Hemisphere. Most nest on remote islands, like this king and these royal penguins on Macquarie Island, 500 miles south of New Zealand.

Feather Detail of Royal Penguin. (bottom left) Photo by Lisle Gwynn. Being flightless, penguins have no need for long flight feathers. Instead, their bodies are covered in almost scale-like feathers no longer than an inch and an half, overlapping one another to create a water-tight shell. The base of these feathers has a small tuft of down, creating an air space that provides insulation from cold temperatures.

Galápagos Penguin and Bubbles. (bottom right) Photo by Angus Fraser. A long stream of bubbles trail behind a Galápagos penguin like a jet stream. While some air escapes from the body feathers, almost all of the air trapped in a penguin's layer of down remains and keeps the bird warm and dry.

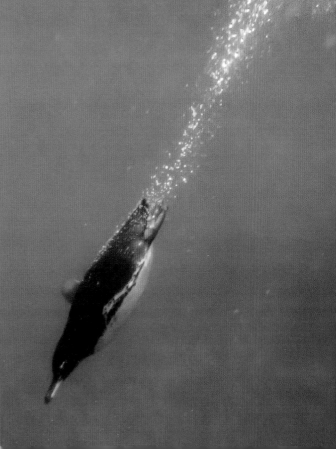

MEET THE PENGUINS

Penguins are perhaps the most universally recognized group of birds in the world. Adopting an upright stance, possessing flippers instead of wings, and conforming to a rather uniform body shape (streamlined) and color pattern (black and white), it is impossible to mistake a penguin for anything else.

Nonetheless, scientists can divide the penguin clan into six major groups, based upon particular characteristics like size, tail structure, or feathering. Three of these major groups have only 1 or 2 members, with the yellow-eyed penguin comprising one, the little and, if classified as its own species, the white-flippered penguin being another, and the emperor and king penguins being the third. Although the little penguin stands only 22cm tall and weighs just 1.5kg, there is no mistaking it as a member of the same clan as its huge cousin, the emperor, which may outweigh the little by 30 times, and be 4 times its size!

King Penguin Eye. Photo by Lisle Gwynn. Penguins hunt by day, locating their prey underwater by sight.

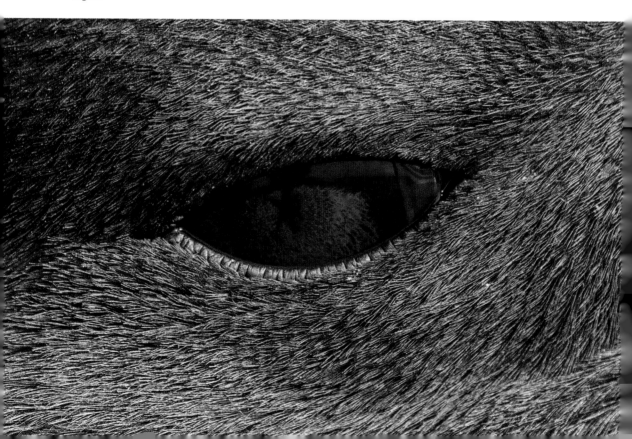

Emperor Penguin Chick.
Photo by Adam Rheborg. There is perhaps no cuter or more cuddly-looking baby bird than the young chick of an emperor penguin. Despite its undeniable cuteness, this chick lives in the coldest and one of the most inhospitable homes on earth, the Antarctic continent.

Adélie Penguin. (top left) The beak of the Adélie appears quite small, as a coating of feathers covers all but the tip. This adaptation may play a small role in conserving heat in its Antarctic environment. **King Penguin.** (top right) The second-largest penguin species, the king penguin is slightly more colorful. Unlike the Antarctic-bound emperor, king penguins are found across the southern ocean on several sub-Antarctic islands. **Emperor Penguin.** (bottom left) Photo by Adam Rheborg. Unlike the Adélie, the emperor penguin's beak is long and bare. Emperors are the largest of all penguins, and one of the most beautiful. **Magellanic Penguin.** (bottom right) You don't have to travel to Antarctic regions to see some penguins. Members of the banded penguin group, like this Magellanic penguin, are found along the coasts of temperate South America.

African Penguin. (top left) Photo by Lisle Gwynn. One wouldn't think lions, jackals, and elephants might share the penguin's land, but historically all 4 could be found in the same area along the southwestern coasts of Africa. Today, only the black-backed jackal remains, while the penguin now faces threats from domestic dogs and cats, motor boats, and even cars, as this species nests along beach-front property. **Galápagos Penguin.** (center) Photo by Angus Fraser. Perhaps the most exciting and rewarding location to see a penguin is the Galápagos Islands, where one may have the opportunity to swim right alongside a Galápagos penguin. Although the waters here are chilly, a tourist inside a wetsuit will be warm and comfortable, and endlessly entertained by these penguins. **Rockhopper Penguin.** (bottom left) The crested penguins, like this rockhopper, are characterized by feathery tufts above or behind their eye. This group is the most diverse, with at least 6 different species, although some ornithologists have divided, or "split," the rockhoppers into 3 different species.

Royal Penguin. (top) Photo by Lisle Gwynn. In silhouette, the royal and the macaroni penguin might look identical, but the lighter-colored black and white chin and face easily distinguish the royal from the other species. **Erect-Crested Penguin.** (center) Photo by Lisle Gwynn. Although erect-crested penguins only breed on a few remote islands off the coast of southern New Zealand, once the nesting season finishes, birds wander widely at sea, some even appearing off the Falkland Islands in the southern Atlantic Ocean. **Yellow-Eyed Penguin.** (bottom) Photo by Donna Salett. New Zealand's yellow-eyed penguin is one of the world's rarest and most threatened penguin species. In addition to habitat loss and predation by domestic animals, the penguins face another new, but completely avoidable foe. Tourists with "selfie sticks" crowd in on the birds for a snapshot of themselves and their "cute friend," stressing the birds and causing nest desertion and abandonment.

Macaroni Penguin. Spaghetti-like strands of feathers characterize the macaroni penguin. It is sometimes confused with the rockhopper, but the macaroni has a more prominent feather tuft and a larger, stronger bill.

THE PENGUIN'S HOME

Penguins are seabirds and are the most highly adapted of all birds for life free of the land. To most, however, it is not the sea and oceans that come to mind when one thinks of penguins, but rather, ice and snow and cold. For some species, notably the Adélie and the emperor, this notion is quite accurate. With the exception of the emperor, who may never touch solid rock or ground its entire life, the 16 or so species must come to land to lay their eggs and raise their young. Exactly where they do so is as diverse as the other 16 species themselves.

Some, like the Humboldt penguin, nest and raise their young in burrows in some

Gentoo Penguin. The penguin's home is the sea, but adults must return to the land to nest. These gentoo penguins are exploding out of the surf in what seems like an exuberant display as they return to their mates and chicks.

Adélie Penguins. (top) Photo by Adam Rheborg. Adélie penguins are one of only two species to nest exclusively on the Antarctic continent. They do not nest on the ice, but instead seek open ground to build their nest. These birds are still waiting for spring and warmer temperatures to clear their nesting grounds of snow. **Rockhopper Penguins.** (bottom) Rockhopper penguins are tough and feisty. When returning to land, they may be tossed and pitched, rolled and slammed, but emerge from the sea unharmed and hop up the sea cliffs to their nests.

of the driest deserts in the world, along the coasts of Peru and Chile. Another lives along the equator, again in a dry, desert habitat, but alongside such tropical species as iguanas, sea turtles, and pelicans. A few nest in tangled rhododendron forests and fairyland landscapes of thick ferns, while most make their land-based home on barren, rocky shores, dotted perhaps with enormous heads of grass.

Adélie Penguins. (following page, top) Penguins must return to the sea to feed. Sometimes entire flocks will bunch up, seemingly unsure of their next move until one brave bird dives in. Within seconds an entire flock may follow. **Chinstrap Penguins.** (following page, center) Once at sea, penguins of every species, like these chinstraps, will take a rest upon icebergs and sheets of ice. **Adélie Penguins.** (following page, bottom) Sometimes hundreds or thousands of penguins may gather on a floating ice sheet. Ice shelters such as these will provide some measure of protection from orcas (killer whales) and leopard seals.

Adélie Penguin. Penguins and icebergs seem to be synonymous, like this Adélie and a wave-carved iceberg.

Bartolóme Island, Galápagos Islands. (top and center) The Galápagos archipelago straddles the equator and the barren, desert-like shores of these volcanic islands and perhaps seems an unlikely home for any cold-loving penguin. Fortunately, the Humboldt Current swoops up from the southern ocean, bathing the islands in cooler water temperatures, thus providing a home for the Galápagos penguin. **Galápagos Penguins.** (bottom and following page) Marine iguanas, most certainly a tropical species of reptile, share the shoreline rocks and cliffs with Galápagos penguins. Nowhere else in the world are species with such radically different living requirements able to share the same landscape.

Galápagos Penguins. (top and bottom) Photos by Angus Fraser. Penguins are seabirds, and the first Europeans to encounter them did not really know if penguins were fish or birds, or a combination of both. Observing Galápagos penguins as they dash and dart and swerve and swoop, catching fast-moving fish effortlessly, one could understand how they might be mistaken for a fish. How could a bird ever move this fast underwater?

Yellow-Eyed Penguin. (top) Photo by Lisle Gwynn. A yellow-eyed penguin rests on a rock ledge along a remote New Zealand island. If nesting, this rock will serve only as a stop-over before the bird walks and hops inland to a proper site. **Fiordland, New Zealand.** (center) Photo by Adam Rheborg. The steep cliffs of the fjords on New Zealand shelter several species of penguin. Here the rare Fiordland penguin makes its home. **Enderby Island and Yellow-Eyed Penguin Habitat.** (bottom left) Photo by Adam Rheborg. Perhaps most strangely of all, yellow-eyed penguins nest in the thick, nearly impenetrable forests and fern meadows of southern New Zealand. The birds may walk for hundreds of yards from the seacoast to reach this nesting habitat.

King Penguin Colony. (top) On some coastlines, like this one at St. Andrew's Bay in South Georgia, the entire landscape is covered by king penguins. The birds nest in open areas, resting their eggs on their webbed feet and forsaking any nest. **Rockhopper Penguin.** (bottom) This rockhopper, high on a hilltop, marched across challenging terrain from one side of an island to the other. Knee-high grass tussocks dotting the landscape made walking difficult for both penguin and photographer, although the penguin had the option of swimming around the island to the other shore. Why it did not is a mystery.

Macaroni Penguin Colony. (top left) A huge colony of macaroni penguins cover an open area of hillside in South Georgia's Cooper's Bay. Like other colonial nesting penguins, the space between nests is defined by the distance one bird can reach when sitting on the eggs.
Macaroni Penguin Colony. (top right) This large nesting colony on South Georgia in the south Atlantic Ocean is typical, situated on a remote, predator-free island amidst spectacular scenery. Many penguin colonies around the world are located in difficult-to-reach landscapes, and here the birds thrive, sometimes with hundreds of thousands of birds at a single site. **Macaroni Penguin Hopping.** (bottom) This macaroni penguin was one of a pair nesting in a large colony of rockhopper penguins on Saunders Island in the Falklands. In this locale, macaroni penguins are rare, and occasionally a lone individual may breed with the similar-looking rockhopper, producing a hybrid.

Humboldt Penguins. Photo by Adam Rheborg. It can be difficult to distinguish the Humboldt penguin from the similar-looking Magellanic, but the more prominent and larger pink patches on the former's face are distinctive. Humboldt penguins ply the cold waters of the Humboldt Current on the west coast of central South America.

BANDED PENGUINS

Four species comprise this group of medium-sized penguins, some also known as "jackass penguins" for their loud, braying call that does indeed sound like a donkey, particularly the noisy African penguin. These 4 species are found farther north and are more readily accessible to human visitation than any other species. The Galápagos penguin sets the record for the most northern, and individuals may on occasion stray north of the equator as they swim about the Galápagos Islands. On the east coast of South America, Magellanic penguins nest in colonies about 1500 miles south of Buenos Aires, and on the west coast of that continent, Humboldt penguins follow the cold ocean currents as far north as Ecuador, just a few hundred miles south of the equator. In Africa, that continent's only penguin species ranges as far north as Namibia, well into the temperate zone.

All of the banded penguins make burrows, hide in crevices, or dig deep scrapes

Humboldt Penguin. Photo by Adam Rheborg.

beneath brush or grass tussocks, which allows these hardy birds to nest in what would otherwise seem to be very non-penguin-like habitats. Safe inside burrows during nesting, Humboldt penguins and some populations of African penguins inhabit some of the driest regions on earth.

Galápagos Penguin. Photo by Adam Rheborg. Quite literally, Galápagos penguins, the smallest of the banded penguins, are "the polar opposite" of one's conception of penguin range and habitat. Instead of ice and frigid temperatures, this penguin lives on dry, rocky shorelines in tropical, indeed equatorial, temperatures in the Galápagos archipelago that straddles the equator.

Magellanic Penguin. Although the most common of the 4 species of banded penguins, Magellanic penguins suffer terrible losses every year off the coasts of South America from oil spills. Mainland colonies are also threatened by predation from feral dogs.

African Penguins. The only species of penguin that nests in southern Africa, African penguins are well-known for their braying call. Over-fishing for sardines and anchovies off the southwest coast of Namibia and South Africa has severely impacted nesting success for this species, as birds must travel farther out to sea to capture prey.

African Penguins. (top left) Look carefully and you will notice the bush hyraxes in the shrubs on either side of this African penguin pair. Human encroachment has changed this penguin's nesting habitat and its inhabitants, which it once shared with the hyrax's distant relative, the African elephant, as well as with African lions, leopards, and South African fur seals. **African Penguins.** (top right) Luxury homes and condominiums line the headlands that surround many of the African penguin's nesting colonies. Considering the remote location of most penguin species' nesting sites, this juxtaposition seems incongruous. Fortunately, the two disparate species—human and penguin—get along just fine! **African Penguin.** (bottom left) Just over 100 years ago, over 1 million African penguins nested along the southern coasts of Africa. Today, the African penguin population hovers at a shocking 50,000 or so birds and, if it continues to decline, Africa's only penguin may be extinct in the wild by 2026. **Humboldt Penguin.** (bottom right) Photo by Adam Rheborg. The least familiar of the 4 banded penguin species, Humboldt penguins are found along the coast and off-shore islands of Peru and northern and central Chile. Threatened by over-fishing, the population estimates for this species range widely, from 12,000 or so to less than one-third that number!

Galápagos Penguin. Photo by Lisle Gwynn. Weighing in at about 5 pounds, the Galápagos penguin is the smallest of the banded penguins. This equatorial species shares the rocky shoreline with colorful Sally lightfoot crabs and marine iguanas.

Galápagos Penguin Molting.
(top) Like all penguins, Galápagos penguins go through a complete feather molt, when old feathers are replaced by a shiny new set. During this time birds remain out of the water as the birds are no longer waterproof and could become chilled in the surprisingly cold water surrounding these tropical islands. **Galápagos Penguin.** (bottom) This bird appears to be flying, as it is completely off the ground, but it was merely captured in mid-jump between rock ledges. Penguins may drop twice their body length from one rock to the next, bouncing slightly as they land, but apparently without suffering injury.

Galápagos Penguins Swimming. (this page) Photos by Adam Rheborg. Unless you are an expert diver and are properly equipped, swimming with penguins isn't an option for most penguin lovers. However, that's not the case in the Galápagos Islands, where the water is warm enough for tourists to tolerate, especially when wearing a wetsuit, and the Galápagos penguins are curious and trusting. Doing so is a real thrill! **Galápagos Penguins Swimming.** (following page) Photos by Angus Fraser.

Magellanic Penguins. (top) The white belly feathers of these Magellanic penguins glow prominently in the late afternoon sun on a Falkland Island beach. Birds rarely linger here, but continue inland to their nesting sites. **African Penguin.** (bottom) Photo by Donna Salett. Banded penguins, like this African penguin, feed closer to shore than other varieties of penguins. This allows nesting pairs to visit their nests and to share incubating and babysitting duties more frequently, although this puts the birds' nesting success at risk if they must roam far from shore to fish.

Magellanic Penguins at Dusk. (top) Penguins normally feed by day, but as the sun sets, Magellanic penguins emerge in numbers from their burrows, where adults and young stand and frequently bray as the light gradually fades to darkness. **Magellanic Penguin.** (bottom) Throughout the day, but particularly at dawn and toward dusk, Magellanic penguins bellow out their distinctive bray. On Sea Lion Island in the Falklands, this honking carries for hundreds of yards, background noise in a windswept landscape.

Adélie Penguin. The size, shape, color, and geographic range of the Adélie penguin cements the conception of what a "typical" penguin must be like. Ironically, of the 17 to 20 species, the Adélie is among the most remote, found only on the continent of Antarctica and surrounding islands and waters.

BRUSH-TAILED PENGUINS

Three diverse species comprise this group of penguins, all characterized by a stiff, bristle-like tail that acts like the third leg of a tripod when the birds stand upright. Climate change has shifted the population dynamics among the brush-tailed penguins, with gentoo penguins, a more temperate species, moving south and sometimes taking over nesting sites of the more cold-tolerant Adélie penguins. In turn, Adélie penguin colonies are declining in the more northern portion of their range, although new colonies may be established farther south on the Antarctic continent. Recently, an enormous new colony of Adélie penguins was discovered off the coast of the Antarctic peninsula, a hopeful sign for this species which was thought to be declining.

The most wide-spread species, the gentoo penguin, nests in the sub-Antarctic across the southern hemisphere. They are one of the most accessible groups for the average penguin lover to observe. Multiple

Chinstrap Penguin. This image of a chinstrap clearly illustrates how this species gets its name. The black line of feathers resembles the strap of a helmet, as this bird surely seems to be wearing.

colonies are found on the Falkland Islands, 600 miles off the southeastern coast of South America, a popular ecotourism destination.

The third species, the chinstrap, achieved worldwide fame and notoriety when two males housed at New York's Central Park Zoo formed a pair bond and began to incubate a rock! Subsequently a fertilized egg from another chinstrap nest was placed into the pair's nest; they incubated the egg, it hatched, and they raised the chick.

On a related note, for several years, a lone male chinstrap penguin staked a nesting site in a gentoo penguin colony in the Falkland Islands, thousands of miles from the nearest chinstrap colony. Each day during the season, the chinstrap pointed its bill toward the sky, cackling its staccato cry in a futile attempt to attract a mate.

Adélie Penguins on Iceberg.
(top and bottom) Everyone taking an ecotourism cruise to Antarctica hopes to see two things: icebergs and penguins. Fortunately, if the weather permits, that hope is usually realized, and often you'll see two together, as Adélies frequnetly use 'bergs as a resting spot.

Gentoo Penguin. The more temperate-loving gentoo penguin is the most colorful of the 3 species, with a bright orange-red beak and feet and vivid face patch. This one seems to be admiring itself in a still pool.

Adélie Penguin Jumping to Sea.
(top right) On snow or ice, adult Adélie penguins are safe from most predators, but the sea is another matter; hungry leopard seals may lurk, unseen, waiting for the unwary. Adélies often clump up together, seeming to gather courage before one bird takes the plunge. Seconds later, the other birds dive in, hopefully confusing a predator with their swirling, speedy numbers. **Adélie Penguin Jumping.**
(center right) An inviting ice shelf may be 5 feet higher than the surrounding sea, but penguins are strong swimmers and can rocket out of the water to land upon a floe. Where they land is another matter, and the collisions that sometimes occur can be quite comical—although harmless for these tough birds. **Adélie Penguin and Chick.** (bottom) Clad in its natal coat of downy feathers, a juvenile Adélie penguin looks nothing like the adult. By the time it goes to sea, however, the young penguin will have molted and replaced the scruffy juvenile plumage with the tuxedo suit of the adult.

Adélie Penguin Covered by Snow. Photo by Adam Rheborg. Severe snowstorms can cover nesting Adélie penguins. This may pose the grave threat of nest abandonment, especially if the birds remain snowbound for several days.

Chinstrap Penguin. (top) Photo by Adam Rheborg. The brush-tailed penguins, like chinstraps, build their nests with small stones. These may be collected along crumbled talus slopes or along the shoreline, or stolen from the nests of other birds temporarily away, gathering or stealing stones of their own! **Chinstrap Penguin Porpoising.** (center) Birds returning to their nest sites are often heavy with newly captured fish, making them easier prey for lurking leopard seals. By porpoising, making repeated leaps from the sea, penguins like this chinstrap not only move faster but also the jumps, and frequent direction changes, may confuse a predator giving chase. **Gentoo Penguin.** (bottom) Photo by Lisle Gwynn. This gentoo appears to be wearing lipstick. Birds in the more northern part of the gentoo's range have a less extensive feather covering around their beaks than those found farther south, like those that nest on the Antarctic peninsula.

Gentoo Penguin Chicks. (top) These two gentoo penguin chicks are well on their way to acquiring their distinctive adult plumage. Gentoos often raise two chicks, but the size and plumage difference between these two would seem to indicate birds from separate nests. **Gentoo Penguin and Whale Bones.** (bottom) A weather-bleached whale skeleton forms an unlikely bed for this gentoo penguin chick. Less than 100 years ago, both whales and penguins were ruthlessly exploited, for both oil and, with penguins, an easy source of food for hungry sailors.

Gentoo Penguin Coming to Shore. A gentoo penguin leaps clear of the water as it races toward shore. What might appear as a shadow is another gentoo swimming in tandem. Seconds later, this penguin would porpoise clear of the water as well. When an incoming wave breaks upon the beach, gentoos often fire out of the water, landing on their feet or belly a few feet in front of the crashing wave before scrambling toward the safety of the beach.

Gentoo Penguin and Magellanic Penguin. Gentoos often nest far from the sea, and nesting birds may waddle hundreds of yards, uphill, to reach their nesting site. At first, this choice of sites may not appear to make sense, but these higher locations often blow clear of snow before the more accessible lowlands, allowing the birds to begin to nest earlier than they otherwise could.

Gentoo Penguins Swimming. (above and following page) Penguins, like these gentoos, are incredibly fast swimmers and extremely agile in the water. Birds can literally do a U-turn within a body's length, and some species can reach over 18mph, several times faster than the fastest human swimmer. With their short, stout flippers, birds "fly" through the water.

Gentoo Plunges into the Ice-Choked Waters Surrounding Its Nesting Colony. This bird may be heading out to sea to fish, where it may make scores of dives to catch nearly its body weight in fish that it will feed to its two hungry chicks.

LARGE PENGUINS

Large size and extremes define this group, with the world's largest living species, the emperor, and the world's longest nesting season, a record held by the kings. The emperor penguin is unique in being the only bird that does not nest on land, as its entire breeding range lies within the Antarctic continent, where it nests upon the ice. Satellite imagery now reveals that some of their ice-bound colonies overlay a solid rock foundation, but nonetheless, the emperor is truly a bird of the ice. This most impressive and striking bird may be more susceptible to a warming climate than any other species of penguin, if the ice sheets that blanket coastal Antarctica melt. Perhaps the bird will adapt, but unless or until it does, melting ice and slush may make it impossible for adult birds to commute to and from their feeding grounds to

King Penguins. The two large penguins are indeed big birds, with king penguins standing nearly a yard tall, and emperor penguins often over a foot taller. Distant ancestors grew even larger, nearly 6 feet tall and over 200 pounds.

Emperor Penguin. Photo by Adam Rheborg. Although king penguins and emperor penguins look very similar, one major difference reflects an important adaptation. The beak of the emperor appears much smaller than the king's, as more of the beak's length is covered by a protective layer of skin and feathers to shield it from the frigid temperatures of Antarctica.

their nesting colonies, and fledgling birds or entire colonies risk dropping into the sea as their ice sheet nesting grounds grow thinner.

The slightly smaller king penguin has a more promising future. Kings are found on several different islands in the sub-Antarctic, where temperatures in summer can be almost balmy. Because of the king's size, a large colony looks absolutely enormous and can cover hundreds of acres, blanketing the landscape in a tapestry of black and white, dotted with the chestnut-colored coats of the immatures. King penguins successfully breed every 3 years, as it takes 16 to 18 months for their single chick to become independent.

> **"King penguins successfully breed every 3 years, as it takes 16 to 18 months for their single chick to become independent."**

King Penguin. (top) Photo by Lisle Gwynn. The beak of the king appears longer than that of the emperor, and the plumage more colorful, with more prominent patches of orange surrounding their dark head. **Emperor Penguin Feet.** (bottom) Photo by Joe Van Os. Although king and emperor penguins look similar, one big difference lies at their feet. Emperor penguins' feathers extend all the way down their legs to their feet, while that area on a king penguin is bare.

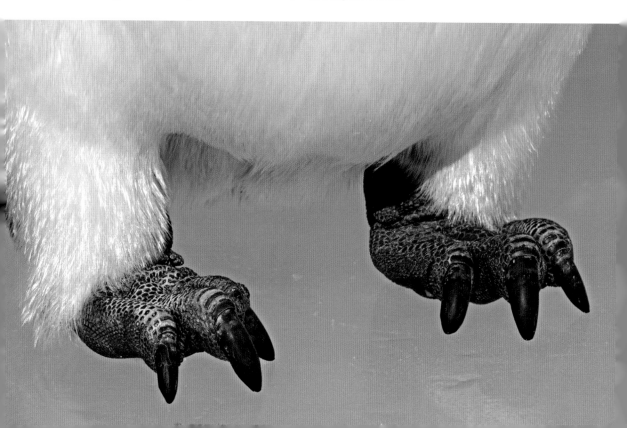

Emperor Penguins. (top) Photo by Joe Van Os. These emperor penguins look like they are standing on a frozen lake, and essentially they are, although in this case the "lake" is the frozen ocean surrounding Antarctica. Emperor penguins are the only birds to nest on ice, and in doing so may spend their entire life without ever touching solid ground. **Emperor Penguins.** (bottom) Photo by Joe Van Os. Almost immediately after their single egg is laid, female emperor penguins head for the sea to replenish themselves and to collect food for their chick. Returning birds will waddle or toboggan for 60 miles or more to reach the colony.

Emperor Penguin Egg. Photo by Joe Van Os. Emperor penguin eggs are huge, measuring about 5 inches long, and are white, at least initially. Over the 60-day incubation, the egg may develop a stain of green or brown. Soon after this egg slipped off its parent's feet, it was doomed, as birds rarely succeed in repositioning the egg.

Emperor Penguins and Chicks. (top left) Photo by Adam Rheborg. Emperor penguins raise only one chick each year, so these two adults are from different nests. Both are waiting for their mates to return from the sea with food for their growing chicks. **Emperor Penguin Chick.** (top right) An emperor penguin chick only a few days old seeks warmth beneath the adult's fold of skin. Females return from the sea soon after their chick hatches; it's likely this adult is the baby's mother. **Emperor Penguin Chicks.** (bottom) Photo by Adam Rheborg. Three different-size emperor chicks gather together to await the return of their parents. The smallest chick is the youngest, and may not survive if spring thaw and melting ice makes an adult's commute to the distant nesting colony impossible.

Emperor Penguin and Juvenile. (top) Photo by Adam Rheborg. This nearly full-grown chick resembles a nearby adult. The chick awaits its final growth of adult feathers before it enters the sea and begins feeding on its own. **Emperor Penguin Chicks.** (center) Photo by Joe Van Os. Few animals can rival the cuteness of an emperor penguin chick. This crèche of half-grown babies appear to be enjoying a balmy day, even though temperatures are likely to be far below freezing! **Emperor Penguin Adult and Chicks.** (bottom) Photo by Adam Rheborg. Chicks of different sizes and ages huddle together near an adult. If temperatures plummet, the chicks will huddle together in a crèche to help maintain warmer body temperature. Birds that do not crèche may not survive inclement weather, which can bring temperatures of –70 Fahrenheit and hurricane-like winds exceeding 120mph!

Emperor Penguin Digestion Rocks. (top) Photo by Adam Rheborg. A pile of stones from an emperor penguin presents a mystery. Scientists do not know why these penguins ingest stones, but various theories propose that the stones may be used as ballast, are consumed accidentally when mistaken for food, or are used as grinding agents, much as those found in a chicken's gizzard. But no one really knows for sure. **Emperor Penguin Chick**. (bottom) Photo by Joe Van Os. Superbly insulated, an emperor penguin chick lying on bare ice is as comfortable as a bird can be. Inside that thick, feathery coat, the bird's body temperature is quite close to that of man.

King Penguin Chick. (top) Visitors to a king penguin breeding colony might believe they are seeing two different species of penguin, as an adult-size juvenile looks completely different from the adult. The furry-looking juveniles remind some of Chewbacca, the fuzzy Wookie from the *Star Wars* movies. **King Penguins on the Beach.** (bottom) King penguins nest on several sub-Antarctic islands, and none nest on the Antarctic continent. During the summer, their nesting grounds are usually snow-free, although unexpected and heavy snowstorms could occur at any time.

King Penguins. (top) Two adult king penguins appear to disagree while a third looks away, noncommittal in the dispute. Before pairing up for the nesting season, disputes such as this are common, as rivals vie for the attention of the opposite sex. **King Penguin and Juvenile.** (bottom) Free from the calorie-burning chore of catching its own food, a nearly full-grown king penguin chick appears much fatter than the two adults nearby. This weight gain is common among many different bird species, and the extra fat is undoubtedly necessary to sustain the chick when it fledges and is forced to hunt for its own meals for the first time.

King Penguin and Elephant Seals. (top) On many islands in the southern oceans, king penguins share the beaches with seals or sea lions. True seals, like these elephant seals, pose little threat to the birds, although an incautious penguin might get crushed should it get in the way of one of these huge pinnipeds when it rolls or moves to the sea. **King Penguin with Wound.** (bottom and following page) Although penguins might scrap and poke at one another in disputes over a nest site or a mate, injuries are rarely serious. Serious wounds like these on a king penguin are likely the result of an attack by a sea lion or a leopard seal. Both prey upon penguins, and the leopard seal often specializes in these birds during the nesting season.

CRESTED PENGUINS

This gaudy, feisty, and decorative group of penguins is the most diverse, with at least 6 different species, but 8 or more are now recognized by most penguin experts. All have, as their name would imply, a plume-like crest that runs above their eye to the back of the head, sometimes forming

Macaroni Penguin or Royal Penguin? (top and bottom) Photos by Lisle Gwynn. In silhouette, the shapes of the macaroni penguin and royal penguin are quite similar.

spaghetti-like strands that hang down about their eyes like a soaking-wet punk rocker, while others have a stiff, almost spiky crest—again, creating something of a punk-rocker look. If you question that analogy, observing their spunky, in-your-face, seemingly cantankerous attitude might convince you. At their nest site, they squall and peck—the penultimate unfriendly neighbor.

My favorite of this group is the rockhopper penguin, now generally split into 3 different species, and they are, to me, the avian equivalent of a football. These birds are tough, and when you watch a flock porpoising through rough seas toward a wave-smashed cliff, you can't imagine anything, especially a knee-high bird, surviving a landing on those shores. Yet they do, with birds popping out of churning white surf, landing on a rock, and hopping toward safety. More often than not, they don't make it, as another wave sweeps in and washes the birds out to sea. The next wave, or one of the others that follow, brings the rockhoppers back to the cliffs once again, and eventually the birds get ahead of the punishing waves and begin what may be a 150-meter climb to their nesting site. These are tough birds!

Distinctive Plumage. (top and bottom) Photos by Adam Rheborg. Both have heavy bills and punk-rocker hairdos, but the gray or white-colored throat of the royal distinguishes it from the black-throated macaroni penguin.

Erect-Crested Penguin. Photo by Lisle Gwynn. Although relatively little is known about this penguin, one fact stands out as truly unique. Erect-crested penguins lay two eggs of quite different sizes. The second egg laid averages over 80 percent larger than the first, with the first egg never surviving and usually removed from the nest by an adult.

Fiordland Crested Penguin and Snares Crested Penguin. Photos by Adam Rheborg. Fiordland (top) and the snares crested penguins (bottom) nest along the coasts of southern New Zealand or surrounding islands. Although both are not common, the Fiordland penguin, nesting on the mainland, is vulnerable to introduced predators like cats and rats. Snares crested penguins only nest on the Snares Islands, where these introduced predators are absent.

Snares Crested Penguins. (top) Photo by Lisle Gwynn. Two snares crested penguins square off in a rare water-based dispute. All crested penguins are feisty, but most aggressive or confrontational behavior occurs around the nest sites, not in water. **Rockhopper Penguin.** (bottom) Photo by Adam Rheborg. Slight differences in the shape of their feathery plumes and significant differences in their geographic distribution has resulted in this species being "split" into 3: the northern, eastern, and southern rockhoppers.

Rockhopper Penguin. (top) Photo by Adam Rheborg. A rockhopper penguin in a typical pose—squawking and complaining to a neighboring bird! Rockhoppers are particularly feisty penguins, constantly squabbling with neighboring black-browed albatrosses and king cormorants at nesting sites in the Falklands. **Erect-Crested Penguins Porpoising.** (bottom) Photo by Lisle Gwynn. A group of erect-crested penguins porpoise toward shore. Laden with a fresh catch for their single chicks, returning penguins are easy prey for predators like New Zealand fur seals. Porpoising confuses pursuing predators.

Snares Crested Penguins. (top) Photo by Lisle Gwynn. Like torpedoes fired into the sea, snares crested penguins leave a long trail of bubbles as they descend, joining other penguins seeking schools of fish. **Royal Penguin.** (bottom) Photo by Lisle Gwynn. Although royal penguins are often soiled by guano from neighboring birds while at their nesting colony, at sea and freshly washed, their colors are vibrant and clean.

Snares Crested Penguins. (following page, top) Photo by Lisle Gwynn. The crested penguins, like this snares crested, could also be called "hopping penguins" for their habit of hopping or jumping up and down the steep ledges where they nest. **Royal Penguin.** (following page, bottom left) Photo by Adam Rheborg. This royal penguin is part of a huge breeding colony on remote Macquarie Island, located between New Zealand and Australia. Like almost all penguins, however, the bird is fearless and curious, and approached the photographer's lens for a closer look. **Rockhopper Penguin Shower.** (following page, bottom right) Returning from the sea, rockhopper penguins on Saunders Island in the Falklands pause at a freshwater seep where they drink and shower, one of the few places in the world where you will see this behavior.

Rockhopper Penguins. (top) One rockhopper completes its jump while another waits in line to follow along. Rockhoppers may nest over 100 yards above the sea, and progress to their nest site in 6-inch-high leaps. When a ledge is too high, rockhoppers gain purchase by using their beak as well, and succeed in climbing rock faces that seem too steep for a bird that is only 20 inches tall. **Rockhopper Claw Marks in Rocks.** (bottom left and right) Stone ledges at favorite access points or steps are often scoured deeply by the sharp claws of rockhopper penguins climbing toward their nest sites. One has to wonder how many thousands of birds had to pass by in order to etch this solid rock.

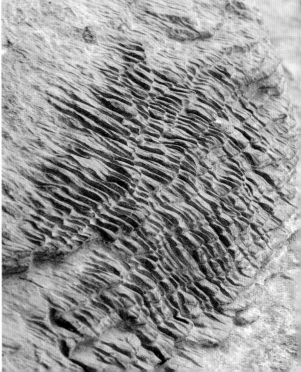

Rockhopper Penguin and Chick. (top left) For the first few weeks after hatching, a rockhopper penguin chick remains beneath the adult's body, where it is safe from predators like skuas, gulls, and caracaras. **Rockhopper Penguin and Chick.** (top right) A hungry rockhopper chick calls and pecks at an adult's beak as it begs for food, prompting the adult to regurgitate a slurry of fish, squid, or krill. Gulls often wait nearby and steal a meal right in front of the penguins' beaks! **Royal Penguins.** (bottom) Photo by Lisle Gwynn. Resembling children in uniform marching home from school, four royal penguins leave the surf and begin their climb to nests far above the beach.

LITTLE AND YELLOW-EYED PENGUINS

The species that originally gave the group its name, the little penguin, has now been "split" or classified into 2 distinct species. The latest species recognized looks similar to the little penguin, which lives in Australia, but the new species has paler flippers and is known as the white-flippered penguin. This species lives only on a few offshore islands in New Zealand.

New Zealand may be the epicenter for penguin evolution, with more species located there and on adjacent islands than anywhere else. One of these residents lives nowhere else, and is, sadly, one of the most endangered penguin species. This is the yellow-eyed penguin, the only member of its group. Colonies that nest on the main island are threatened by habitat loss, human disruption, and predation from feral introduced species, including rats and cats.

Little Penguins. (top) Photo by Adam Rheborg. Little penguins are also known as blue penguins because of their dorsal coloration. This and the closely-related white-flippered penguin are the two smallest species, and weighing just 3 pounds or so they are 3 percent the weight of their largest cousin, the emperor penguin. **Little Penguins and Car.** (bottom) Photo by Adam Rheborg. Although little penguins are diurnal, or day-active, they return to their nesting colonies after sunset. Consequently, in habitat that they share with humans, little penguins are exposed to many dangers, including traffic as the birds cross roads to reach their nests.

Yellow-Eyed Penguin. (top) Photo by Donna Salett. There are less than 2,000 breeding pairs of yellow-eyed penguins, endemic to New Zealand and surrounding islands. It is often considered the rarest and most endangered species, although in El Niño years, the small Galápagos penguin's population can drop by over 80 percent, making it the rarest penguin until its population rebounds in subsequent years. **Yellow-Eyed Penguins.** (bottom left and right) Photos by Lisle Gwynn. Thick ferns and twisted, gnarled trees in a dense temperate forest are probably not what comes to mind when one envisions a penguin's nesting grounds, but that's precisely the habitat yellow-eyed penguins select as they build their nests. Unlike other penguins that frequently nest just out of pecking reach of each other, yellow-eyed penguins are solitary nesters, with nests separated by 50 yards or more.

Yellow-Eyed Penguins. (top and center) Photos by Lisle Gwynn. No other species of penguin has the distinctive eye color that gives the yellow-eyed penguin its name. Adults, juveniles, and birds in molt can always be identified by this unique feature. **Yellow-Eyed Penguin.** (bottom) Photo by Lisle Gwynn. The brilliantly colored eye of this yellow-eyed penguin blazes through a veil of water as it surfaces after a dive. In addition to their striking eyes, this species has perhaps the most colorful beak of all the penguin species.

THE NESTING SEASON

The natural world is, of course, filled with wonderful examples of adaptation, some so extreme as to almost defy belief. The penguins are no exception and, in fact, the nesting behavior of one species in particular represents perhaps the most extreme example of persistence and fortitude of any vertebrate. Imagine fasting for months, suffering temperatures far below 0 degrees, and hurricane-like winds, while incubating an egg on your webbed feet in the dead of an Antarctic winter! That is what the emperor penguin endures, as it hunkers over its egg, shuffling in a circle that ensures that birds share, somewhat equally, the relative comfort of the center of a massive huddle of other birds and the brunt of the wind and cold when rotated along to the outer ring.

Other penguins enjoy a less extreme lifestyle. Some, particularly the banded penguin group, often nest in burrows, while brush-tailed penguins build little pyramid-shaped nests of stone or bone or strands of algae. Kings, like their cousin the emperor, make no nest at all, but incubate their single egg while it rests on their feet, protected from the elements and warmed by a fold of skin. None lay more than two eggs, and in some cases, the first egg is just

King Penguin Colors. Photo by Lisle Gwynn. While the snappy colors of this king penguin might be pleasing to the human eye, to another penguin the vibrancy of color is likely an indicator of good health and vigor, important qualities for a prospective mate.

a fraction of the size of the second egg and is later ejected from the nest. Several species of penguins will raise two chicks, if food resources allow it.

King Penguins Mating. (top left) An attempt at mating puts the male, on top, in a precarious position, and a successful mating occurs in just a third of all attempts. A third of the time, the male simply loses his balance and falls off, while the remaining attempts simply miss the mark. Persistence, however, eventually pays off.
African Penguins. (top right) Photo by Donna Salett. Penguins, like these African penguins, reinforce their pair-bond throughout the nesting period, conducting head-bowing ceremonies or mutual "crowing," as well as periodic grooming sessions. **King Penguin Colony.** (bottom) Most penguin species breed in colonies, but few are as spectacular as that of the king penguins, where hundreds of thousands of birds may gather, literally covering all available space on their nesting grounds.

Little Penguin. (top) Photo by Cindy Marple. Even burrowing species, like this little penguin in Australia, breed in colonies, although the space between one nest and another might be several yards. In these species, colonies may form because the habitat is suitable for burrows, rather than for mutual protection from predators. **Magellanic Penguin and Chick.** (bottom) The burrow of a Magellanic penguin is a safe refuge for a young chick that would otherwise be vulnerable to the attack of skuas, gulls, or caracaras. If a threat appears, both the adult and chick can scoot into their nest.

Emperor Penguin and Chick. (top) Photo by Adam Rheborg. Emperor penguins nest on bare ice, and one of the adults must remain with their single chick until it is old enough to safely join other chicks in a crèche. While nest predators are few, Antarctic skuas, a hawk-like gull, will steal eggs or chicks from any unwary parents. **Emperor Penguin Coated in Snow**. (bottom) Photo by Adam Rheborg. A coating of snow packs the face of this incubating emperor penguin. Once the egg is laid, the female returns to the sea to feed. The male, who may have been at the nest site for a month or more, will be the sole incubator for the next 64 days, resting the egg on his webbed feet the entire time.

Adélie Penguins in Snow. Photo by Adam Rheborg. Snowstorms can affect nesting. A heavy snowfall may cover nesting Adélie penguins completely, with only their heads poking above the snow. Birds will endure these conditions for as long as 3 days; but any longer, and the adults will desert the nest.

Gentoo Penguin. A gentoo penguin follows
a well-worn trail back to its nest. Some species
relieve their partners and exchange parenting
duties every few hours, but species that feed
farther out to sea may be absent for several days.

Gentoo Penguin Colony. (top) Gentoo penguins often nest on hilltops and plateaus high above the sea, requiring an arduous waddle to reach their nests. While this might seem to be a poor choice for a nesting site and a waste of energy, these high areas are usually the first spots to be clear of snow. For penguins, the earlier a bird can begin nesting, the better the chances that the chicks will survive, as food becomes more plentiful as fledging time approaches. **Chinstrap Penguin Colony.** (bottom) A male chinstrap penguin may claim a nesting site before the snow melts. While females may pair with the same male in subsequent years, her choice is dictated by the nest site, not the male who owns it! Perhaps if a male penguin could, he'd also drive a nice car!

King Penguin. (top) Photo by Cindy Marple. Nest sites can be far from water. Walking through snow can be difficult, and king penguins and other species will often toboggan along on their bellies, propelling themselves forward with flicks of their flippers and push-offs with their clawed toes. **Adélie Penguin with Stone.** (bottom) Brush-tailed penguins like this Adélie build a nest of stones, bones, and even dried parts of dead penguins. They are not above stealing, either. Any unattended nest is likely to be looted by neighboring penguins.

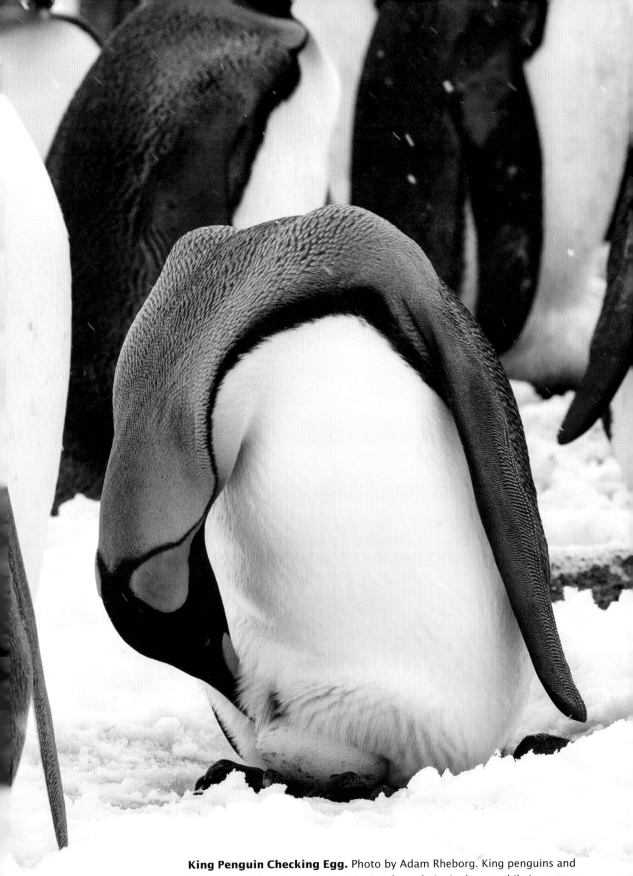

King Penguin Checking Egg. Photo by Adam Rheborg. King penguins and their cousin, the emperor penguin, incubate their single egg while it rests on their webbed feet. The transfer of a fresh egg from the female to the male is awkward, and many eggs are lost when an egg rolls clear of either parent's feet.

Adélie Penguin with Stone. (top) Nests are continually being repaired or rebuilt, as stones come and go with the thefts rampant in the Adélie penguin colony. The presentation of stones, even after the chicks have hatched, may be another pair-bond reinforcement ceremony. **Adélie Penguin with Brood Pouch.** (bottom) The insulation of a penguin's feathers would prevent an egg from staying warm, but instead, the egg is nestled in a featherless, cup-like depression called a brood pouch, where the egg is pressed against bare skin close to warm blood vessels. **King Penguin Colony.** (following page) Male king penguins patiently incubate their single egg while their mate goes fishing. She will return in 18 or 19 days to relieve the male, giving him the first chance in months for a fresh meal.

King Penguin Chick.
(previous page) Although
quite large, this king
penguin chick is still
too small to leave the
protection of a parent.
Unattended chicks are
vulnerable to skuas and
giant petrels and, as here
in the Falklands, striated
caracaras, a very bold
and efficient predator.

King Penguin and Egg.
(top) Lifting his belly
fold, a male king penguin
rolls the egg resting on
his webbed feet. **King
Penguin Chick.** (bottom)
A recently hatched king
penguin chick peers out
at the world. Penguin
chicks are semi-altricial,
as they are covered with
a thin coat of downy
feathers, but are com-
pletely dependent upon
their parents to provide
food. True altricial babies
are hatched featherless,
while the opposite,
precocial hatchlings, are
feathered and can feed
by themselves.

Striated Caracara and Egg. (top) A pair of striated caracaras have successfully harried a rockhopper penguin and stolen an egg, while other rockhoppers squawk their displeasure. At some colonies in the Falklands, caracaras are having significant impact upon the penguin colonies. **Emperor Penguins.** (bottom) Photo by Adam Rheborg. This emperor penguin chick is getting a bit too big for its parent's brood pouch. Soon, it will join other chicks and form a crèche, where they'll gather, awaiting their parents' return with their next meal.

Gentoo Penguin and Chick. (top right) This young gentoo penguin chick still has its "egg tooth," a tooth-like protrusion at the top of its beak that the chick uses to cut, chip, and saw its way free of its shell. The adult doesn't assist in this, though on occasion, may attempt to feed the chick, dropping krill onto the open shell. **Gentoo Chicks and Adult.** (bottom left) Gentoo penguins lay two eggs, and in good years both chicks will survive. Eggs are laid, on average, about 3 days apart, but the first egg is only partially incubated, ensuring that both eggs will hatch within a day or so of each other. **Gentoo Chick Feeding.** (bottom right) Penguins do not carry food back to the nest in their beaks. Instead, the fish, krill, or squid is swallowed, and later regurgitated directly into the chick's mouth after a bit of beak-pecking/begging on the chick's part.

King Penguin Chicks. (top left) It would be easy to believe that these big brown birds are a different species of penguin, as king penguin chicks are clad in a furry-looking coat of brown feathers. Fat from feeding, they can be larger than their parents. **King Penguin Chicks.** (top right) Two large king penguin chicks wander off from the collective safety of the crèche. Soon, they will begin their molt to their striking adult plumage. **King Penguin and a Mob of Chicks.** (below) Photo by Adam Rheborg. How in the world does a mother king penguin find her own chick in a huge crèche? They do so by voice, and although the whistles and trumpets of a king penguin all sound alike to human ears, the birds recognize subtle differences as unique as each human's face or voice is to us.

Gentoo Penguin Chick Chase. (top) When an adult returns to a crèche with food, the chase is on. The parent's chick, along with any other hungry youngsters, immediately mob the adult, prompting the bird to run off. The chicks follow, and after a few laps around the colony, only her chicks remain. **Gentoo Penguin Feeding Chicks.** (center) Two chicks struggle to catch Mom's meal. Normally, one chick gives up first, which allows the adult gentoo to feed one chick at a time. In this case, two nearly equally sized chicks tired the adult out first! **African Penguin and Chicks.** (bottom right) These two African penguin chicks are large enough to remain outside their burrow, at least until danger threatens. Pecking at the adult's beak will stimulate the adult into relinquishing a meal.

MOLTING

All birds molt, shedding their coat of old feathers for new, fresh plumage. In some bird species, this happens in barely noticed stages, allowing the bird to continue with its normal activity relatively unimpeded. Others lose their flight feathers and breeding colors relatively quickly. Ducks are a prime example, and in late summer, the sexes look drab and similar, and are flightless until their new plumage grows in.

Penguins molt in a somewhat similar way, and perhaps because the change in penguins

King Penguin Pair. (left) At the start of the breeding season, king penguins are resplendent in plumage designed to catch the attention of the opposite sex. In the months that follow, as birds lay eggs, raise chicks, and commute back and forth from fishing grounds to nest sites, their energy is sapped and their plumage suffers. **King Penguin Preening.** (right) The molt may begin with an itch, with a king penguin picking at some feathers that begins the long process of renewal.

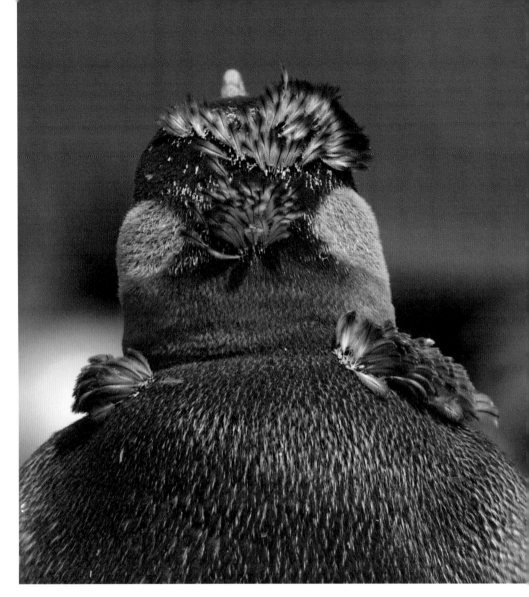

King Penguin Molt. The first old feathers push out, giving this king penguin a lumpy, warty-looking appearance. As you can see, all of the other feathers are dull, lacking the shiny gloss one expects in a penguin's appearance.

is even more dramatic, and limiting, their variety of molt is known as a catastrophic molt. During this molt, penguins look ruffled and unkempt, and more importantly, they are no longer waterproof. Penguins during this catastrophic molt must remain on land, and those entering the water before their new feathers grow in risk hypothermia if cold water reaches their skin. This molting period is the most stressful period in a penguin's life, as the birds must fast while on land, and must use their stored-up fat not only to exist, but also to grow new feathers. Fat alone isn't sufficient to do both, and in the final days of the molt, as new feathers replace the old, penguins must use energy derived from consuming muscle, further stressing the birds. After a few weeks, or two months in the largest penguins, a new coat of sleek, bright feathers emerges, and the birds are once again equipped for a life in the cold seas. Hungry and weak, the penguins are ready for their first meal in several weeks!

Chinstrap Penguin. (top) Photo by Adam Rheborg. The first of its old feathers pushing out, this chinstrap's headdress resembles an Indian war bonnet.

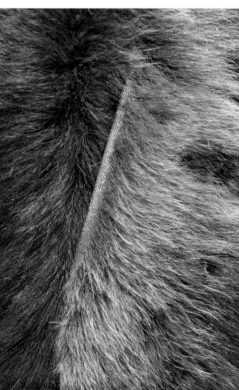

King Penguin Chick. (bottom left) At some point, adult king penguins stop feeding their now enormous chick. This juvenile will begin the molt that will replace its fur-like feathers with the feathers and coloration of an adult. **King Penguin Wing.** (bottom right) The juvenile feathers of a wooly-looking king penguin resemble fur more than feathers.

King Penguin Chick. (top) Photo by Adam Rheborg. The Falkland Islands climate can be rather pleasant and un-penguin-like. Nonetheless, these are sub-Antarctic islands, and on any given day, a juvenile king penguin could experience snow, hail, and freezing temperatures. Fluffy feathers prepare this chick for those extremes.

Gentoo Penguin in Molt. (bottom) Photo by Adam Rheborg. At first glance, a molting penguin might be unrecognizable, as old feathers project outward in unexpected angles. Depending upon the species, this molting period lasts from 2 to 5 weeks. During that time, the bird will not feed and will try to remain as inactive as possible as it grows new feathers, consuming vital energy as it does so.

King Penguin Chick. (top) Photo by Adam Rheborg. Just beginning to lose its juvenile coat of fluffy brown feathers, this king penguin resembles a school kid who just tramped though a puddle. Although unattractive to some, tourists often find these birds in mid-molt cute and certainly funny-looking. **King Penguin Molt.** (bottom) Photo by Adam Rheborg. The vibrant colors and striking pattern of an adult king penguin begins to reveal itself as this bird loses its juvenile plumage. Soon, it will resemble an adult, though it may be several years before this juvenile assumes nesting duties of its own.

King Penguin Chick in Water. (top) Photo by Lisle Gwynn. Clad in a mix of juvenile brown fluff and adult plumage, this king penguin may risk hypothermia, as the juvenile feathers are not waterproof. Only when all of the feathers are replaced will this bird be insulated from the frigid waters.

King Penguin Bathing. (center) After breeding, penguins return to the water for a massive feeding frenzy to put on fat and muscle that will be vital to surviving the weeks-long molting process. This king may be enjoying its last swim before being landlocked and hungry for nearly a month. **King Penguins Molting.** (bottom) Molting king penguins gather outside their nesting colony to molt. To save energy and survive the fast, penguins are inactive. Clumped in groups, they can stay warm if inclement weather swoops down onto the colony.

King Penguins Molting. (top) In varying stages of their molt, these king penguins turn their backs to the wind in order to stay warm. Their sometimes quarrelsome nature is subdued during this critical time. **King Penguin Molt.** (center) This king penguin has nearly completed the molting process, and only an odd-looking cluster of feathers remain, giving the bird a hunchback appearance. **King Penguin Colony.** (bottom) At the end of the nesting season, a king penguin colony can be as quiet as it will ever be, as juveniles transform into adults and molting adults replace old feathers with new. This is the most stressful period in a penguin's life, and birds lacking sufficient fat reserves will die.

King Penguin Gathering. (following page) Standing motionless on the edge of a stream, king penguins conserve energy while old feathers form windrows in the lazy current. On windy days at the peak of the molting, the air can be filled with drifting feathers, creating the appearance of a feathery snowstorm.

PREDATORS

Standing on the edge of a king penguin colony of a hundred thousand birds, it's easy to believe that these birds must be invulnerable, safe from any predator. Of course, the reverse is true, as any animal that numerous must inevitably attract something that would eat it. And that is the case with penguins, who face a limited but nevertheless formidable set of predators.

In truth, however, penguins have co-existed with every one of their predators since the first penguin lost its ability to fly and became a creature of the sea. Seals, whales, and sharks hunted these sea-going birds, while skuas, gulls, caracaras, sheathbills, and in bygone days, giant eagles and more preyed upon the birds on land, especially during nesting. Yet penguins flourished,

and at one time, this group of birds was the dominant warm-blooded predator in the seas. As seals evolved, they replaced the penguins at this top position, but still the penguins flourished.

It was only with the arrival of man that penguins faced their most serious threat, one that continues even to this day. In the past, penguins were exploited as a source of fresh meat and eggs, just as their namesake, the great auk, was exploited and driven to extinction. Far more wasteful and horrific, however, was the carnage created as penguins were herded and marched to huge boiling pots, where they were rendered to oil, wiping out entire colonies in the process. Man has proven to be the penguin's greatest predator and one that still poses a threat today.

King Penguin and Boiling Pots. Photo by Lisle Gwynn. Two enormous boilers used for processing penguins are now surrounded by a thriving king penguin colony. At some locations throughout the southern oceans, processors like this wiped out entire colonies of penguins, as well as Antarctic and southern fur seals.

King Penguin and Antarctic Fur Seal. (top) On land, most of the time, a king penguin is safe from an Antarctic fur seal, as the bulls at this time are intent on finding mates. Nonetheless, occasionally, fur seals kill penguins as if for sport, and others actively prey upon birds swimming to and from shore. **Gentoo Penguin and Boiling Pot.** (bottom) A gentoo penguin appears to be assessing a now-abandoned boiling pot in the Falkland Islands. It is hard to imagine today that less than 100 years ago, penguins would be pushed and herded to slaughter, their fats and oil rendered in these ghastly pots.

King Penguin and Antarctic Fur Seal. (center) A bull Antarctic fur seal blocks the path of king penguins on their way to their nesting sites. Fur seals are not true seals, but belong to the group of "eared" seals, the sea lions. True seals, like the elephant seal, lack external ears, and differ also from the sea lions in that their rear flippers cannot rotate forward. Sea lion flippers do, allowing a sea lion to gallop across a beach fast enough to catch a scurrying penguin. **King Penguin Porpoising.** (bottom right) Porpoising, named for the same behavior shared by true porpoises and dolphins, is the fastest way to move through water. King penguins rarely porpoise, as the speed required to clear the water costs valuable energy. This behavior, often used to elude predators, is practiced often by the smaller penguin species.

King Penguin and Elephant Seal Pup. (previous page, top) Photo by Lisle Gwynn. This king penguin expresses its annoyance at a harmless elephant seal calf on its beach. Bull elephant seals, fighting or contesting a harem, may crush unwary penguins as these huge, 18-foot-long monsters lumber with surprising speed down a beach.

Gentoo Penguin and Leopard Seal. (top right) Photo by Adam Rheborg. Leopard seals often patrol the shorelines of penguin colonies. Weak and hungry after the molting period, penguins like this gentoo are an easy catch. **Injured King Penguin.** (bottom right) This king penguin escaped the jaws of a leopard seal but is still not safe, as the injured bird may be attacked by giant petrels. While this wound looks quite serious, I've seen healthy penguins with old scars that crossed their entire bellies.

Leopard Seal. (top) Public enemy number one for penguins is the leopard seal. They are big seals with long, tapered bodies and a flexible, almost snake-like neck. They remind me of a dragon or the extinct plesiosaur, a marine reptile from the dinosaur era. **Orca.** (center) Various populations of orcas, or kller whales, specialize in different prey. In Antarctica, some orcas have learned to swim as a group toward an ice floe, creating a bow wave that swoops over the ice and dumps a penguin or seal into the sea. A penguin's best defense when pursued is to swim erratically and to porpoise, possibly throwing off the predator. **Giant Petrel Feeding on Penguin.** (bottom) Sometimes called the vulture of the sea, giant petrels are not only scavengers, but are active and fierce predators as well. Fights over possession of a kill are common.

Giant Petrels Fighting over Kill. (top) Several giant petrels spotted an injured king penguin and attacked, and the end was quick for the injured bird. Skuas, gulls, and sheathbills will scavenge what little leftovers remain. **Striated Caracara and Rockhopper Penguin.** (center) On Saunders Island in the Falklands, I watched 4 or 5 striated caracaras repeatedly raid a rockhopper penguin colony. Heavy predation can severely impact a penguin colony, and few chicks survived that season. **Striated Caracara Carrying Penguin Chick.** (bottom) After snatching a young penguin, striated caracaras would frequently drop the bird onto the rocks below, abandoning the kill and going after another. Young rockhopper penguins in crèches were most vulnerable, as adult penguins were rarely close enough to offer effective protection.

Turkey Vulture Carrying Rockhopper Penguin Chick. (top) Turkey vultures normally scavenge, and it is likely that this vulture is merely carrying off a rockhopper penguin chick killed by a caracara. **Gentoo Penguin Driving Off a Skua.** (bottom) Photo by Cindy Marple. An adult gentoo penguin drives off a Falkland skua intent on stealing the chicks' meal. Skuas will rob nests of eggs and small chicks, but chicks of this size are safe from predation by this bird. Skuas do harass the birds, however, and will snatch regurgitated fish intended for their hungry chick.

Skua Stealing a Penguin Egg. (top) Photo by Donna Salett. Skuas are masters at egg theft. In some years, nearly 20 percent of the eggs of Adélie penguins are taken by skuas at certain colonies.

King Penguin in Surf. (bottom) A king penguin prepares to join others as they head to sea. Traveling in a group lessens the chance of predation, at least for each individual bird. When pursued, fleeing penguins darting in multiple directions may confuse a predator, increasing the chances of every penguin surviving an attack.

THE PENGUIN'S FUTURE

Today, of course, penguins are no longer boiled alive for oil, but ironically, entire colonies risk extinction when oil spills, from off-shore rigs, illegal dumping, or ship wrecks, drift in vast, deadly pools across the seas and onto nesting beaches. In the 1980s and '90s, as many as 20,000 Magellanic penguins died each year from oil spills. In 2011, a massive oil spill off a remote southern mid-Atlantic island killed thousands of northern rockhopper penguins, nearly wiping out both the colony and the species. In the Falklands, home to some of the most human-accessible colonies of penguins, including king, gentoo, rockhopper, and

Gentoo Penguin Colony. Climate change and increasing temperatures are most noticeable in the polar regions, which may result in shifts in krill or fish population hubs. Nesting gentoo penguins, which normally feed far from shore, may need to travel even further as they hunt for fish and krill.

Macaroni Penguin and Chick. Once ruthlessly exploited, penguins in the Falkland Islands now enjoy complete protection and are the foundation for a thriving ecotourism industry. Species that may have nested on these islands long ago are now returning, and today small groups of macaroni penguins nest among scattered rockhopper penguin colonies.

Magellanic, the prospect of off-shore oil drilling casts a dark shadow over these birds' future.

Over-fishing looms as a serious threat, and African penguins have suffered greatly as anchovies, sardines, and similar fish have been depleted. Since 2004, this species has been reduced by 70 percent. A century or so ago, millions of these penguins nested along the southern shores of Africa, but today, this endangered species numbers less than 50,000. African penguins, like all the banded penguins, are near-shore feeders, and when fish stocks are depleted around their nesting colonies, birds must travel farther out to sea, requiring longer periods away from growing chicks or mates still guarding nests. Nest desertion may result, as hungry birds must choose between remaining at a nest for a mate that may never return, or going off to fish and live to breed another day. As close-at-hand schools of anchovies and sardines disappear, desertion becomes more common. Over-fishing, including the relatively recent harvesting of krill for feeding captive-raised fish for human consumption, threatens not only penguins but all fish-feeding animals, ourselves included.

Not long after the extinction of the dinosaurs, some 63 million years ago, the penguin clan proliferated into nearly 3 times the number of species that exist today. With the evolution of seals and sea lions, the penguins dominance of the sea faded, resulting in the 17 to 20 recognized species today. Ironically, as the great whales, the blue, fin, sei, Bryd's, and others were exploited nearly to extinction, fur seals, which also suffered massively, have rebounded, and today their burgeoning numbers threaten penguin survival once more. Fur seal breeding season

Galápagos Penguin. Photo by Angus Fraser. Periodic El Niño events affect the food supply for Galápagos penguins. This species forages close to shore, and shifting fish populations can severely impact nesting success and even adult survival.

King Penguins and Tourists. King penguins and a small group of ecotourists meander along the extensive flatlands on South Georgia, one of the top penguin-viewing areas in the world.

coincides with the penguins', and these aggressive mammals may soon block access to the beaches and the nesting grounds of birds in several locations in the southern hemisphere.

A changing climate may also affect the penguins. El Niño, the warm-water phenomenon that swoops west to east across the Pacific, impacts Humboldt and Galápagos penguins, as sea temperature rises and the nutrient-rich cold water currents that these penguins depend upon are deflected. Nutrient-rich water nourishes plankton, which is the foundation of the food chain, but warm waters suppress the plankton and the food chain collapses, taking down with it the sea lions, boobies, pelicans, and penguins that fish the cold waters off the Galápagos and western South America. At one time only a once-a-decade phenomenon, El Niños are occurring with greater frequency, and the effect can be devastating. During the El Niño of 1982–83, the Galápagos penguin was particularly hard hit, and over 75 percent of the population died from starvation. Today, it is estimated

> "During the El Niño of 1982–83, the Galápagos penguin was particularly hard hit, and over 75 percent of the population died from starvation."

that there are fewer than 800 breeding pairs of Galápagos penguins, with a total population between 3,000 and 8,000 birds—an enormous discrepancy in guesstimates, but either way, it makes this bird the world's most endangered penguin.

A warmer climate may generate devastating snowstorms that bury nesting Adélie penguins, may thin the ice sheets for the colonies of emperor penguins, and reduce the total winter ice sheets necessary for the production of krill, a major food source for many penguins. And, along with everything else, there lies another insidious threat that we are only beginning to understand, and that is plastics. Seventeen million tons of plastic are dumped into the ocean every year, and although most plastics degrade into sand-grained size bits and pieces, or even into nano-particles, plastics do not go away. Instead they persist, at the least, for hundreds of years, and these plastic particles are finding their way into the food chain. The effect they may have on the future of life in the ocean is unclear, but the potential is frightening.

Today, however, we can at least take heart in the knowledge that penguins are not exploited for their meat, eggs, or oil. Most colonies are protected, some formally, others strictly so, and previously decimated colonies have in many cases seen a rebirth, with birds returning to historic nesting sites. Besides their intrinsic value as fellow

King Penguins. (below) Even on remote south Atlantic islands, penguins, like these kings, are not immune from outside forces. A warming climate, an oil spill, over-fishing, and myriad other factors far removed from the penguins themselves can have devastating effects. **Gentoo Penguin.** (following page) The vibrant color in the beak and feet of this gentoo penguin testify to a healthy and well-fed bird. This one was returning to its nesting colony after a daylong fishing trip at sea.

Gentoo Penguin. Let's hope that this gentoo penguin, slipping off an ice shelf, is not suggestive of the penguins' future!

denizens of our planet, ecotourism has given many species a tangible dollar value. Tourism in the Falklands is largely penguin based, as it is the Galápagos Islands, though to a somewhat lesser extent. Who would cruise to Antarctica for icebergs alone, if not for the expectation of seeing penguins, too?

For now, the penguin world is stable, though more species are threatened or endangered than are not, but virtually every species, every population, balances upon a razor's edge, where any number of calamities could create an entirely different picture. We can only hope that these

> "For now, the penguin world is stable, though more species are threatened or endangered than are not . . ."

incredible creatures, perhaps the toughest and most adaptable of any group of birds, will always continue to inhabit the world's roughest and most remote regions, and give pleasure to all who have the good fortune to encounter them.

Penguin Tracks. As they have for millions of years, penguins march across a sandy beach. While an incoming wave will wash away these tracks, let's hope that penguins will continue to inhabit the world's wildest places, environmental "canaries in the coal mine" for a healthy planet.

INDEX

Big Cats in the Wild

Joe McDonald's book teaches you everything you want to know about the habits and habitats of the world's most powerful and majestic big cats. $24.95 list, 7x10, 128p, 220 color images, index, order no. 2172.

Polar Bears in the Wild

A VISUAL ESSAY OF AN ENDANGERED SPECIES

Joe and Mary Ann McDonald's polar bear images and Joe's stories of the bears' survival will educate and inspire. $24.95 list, 7x10, 128p, 180 color images, index, order no. 2179.

Bald Eagles in the Wild

A VISUAL ESSAY OF AMERICA'S NATIONAL BIRD

Jeff Rich presents stunning images of America's national bird and teaches readers about its daily existence and habitat. $24.95 list, 7x10, 128p, 250 color images, index, order no. 2175.

Dogs 500 POOCH PORTRAITS

TO BRIGHTEN YOUR DAY

Lighten your mood and boost your optimism with these sweet and silly images of beautiful dogs and adorable puppies. $19.95 list, 7x10, 128p, 500 color images, index, order no. 2177.

Owls in the Wild

A VISUAL ESSAY

Rob Palmer shares some of his favorite owl images, complete with interesting stories about these birds. $24.95 list, 7x10, 128p, 180 color images, index, order no. 2178.

The Sun

IMAGES FROM SPACE

Travel to the center of our solar system and explore the awesome beauty of the star that fuels life on earth. $24.95 list, 7x10, 128p, 180 color images, index, order no. 2180.

Trees in Black & White

Follow acclaimed landscape photographer Tony Howell around the world in search of his favorite photo subject: beautiful trees of all shapes and sizes. $24.95 list, 7x10, 128p, 180 images, index, order no. 2181.

Wicked Weather

A VISUAL ESSAY OF EXTREME STORMS

Warren Faidley's incredible images depict nature's fury, and his stories detail the weather patterns on each shoot. $24.95 list, 7x10, 128p, 190 color images, index, order no. 2184.

The Frog Whisperer

PORTRAITS AND STORIES

Tom and Lisa Cuchara's book features fun and captivating frog portraits that will delight amphibian lovers. $24.95 list, 7x10, 128p, 350 color images, index, order no. 2185.

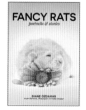

Fancy Rats

PORTRAITS & STORIES

Diane Özdamar shows you the sweet and snuggly side of rats—and stories that reveal their funny personalities. $24.95 list, 7x10, 128p, 200 color images, index, order no. 2186.